CLIMATE ALARM REALITY CHECK

WHAT YOU HAVEN'T BEEN TOLD

For Open Minded Readers

ANDREW L. URBAN

Published by:
Wilkinson Publishing Pty Ltd
ACN 006 042 173
PO Box 24135
Melbourne, Vic 3001
Ph: 03 9654 5446

enquiries@wilkinsonpublishing.com.au
www.wilkinsonpublishing.com.au

Title: Climate Alarm Reality Check

ISBN: 9781922810991

A catalogue record of this book is available from the National Library of Australia.

Design by Spike Creative Pty Ltd
Ph: (03) 9427 9500
spikecreative.com.au

Printed and bound in Australia by Ligare Book Printers.

"It ain't what you don't know that gets you into trouble. It's what you know for sure that just ain't so."

–Mark Twain

The author wishes to gratefully acknowledge Louise Keller for her valuable insights in the preparation of the manuscript.

CONTENTS

This book is dedicated to the panic stricken 12 year old schoolgirl I saw sobbing on the TV news fearing the world will soon end because of 'climate change' – and all her fellow schoolchildren around Australia.

"It is easier for people to believe lies than to convince them that they have been lied to."

–Mark Twain

KEY POINTS

(These points are supported by citations in the text.)

As is well known, the Earth's climate has been changing since before man came along and burnt fossil fuels. Those changes were not driven by carbon dioxide.[1]

- humans generate 0.012% of all carbon dioxide in the atmosphere;
- carbon dioxide makes up 0.04% of the atmosphere;
- contrary to the ruling orthodoxy, scientists do not regard human emissions as drivers of dangerous climate warming;
- the climate threat is not an existential one, even in its most alarming hypothetical incarnations;
- on February 23, 2017, a petition was sent to then US President Trump, signed by some 300 scientists, warning that the push to curtail carbon dioxide threatens to exacerbate poverty without improving the environment;
- climate science has demonstrated noble cause corruption, where the ends justify the means;
- *"Climate policy has almost nothing to do anymore with environmental protection, [but] with the distribution of the world's resources"*– senior IPCC official;
- the media has failed to subject the IPCC to the scrutiny applied to other organisations;

1 The climate has changed for 4.5 billion years via long term, medium term, short term cycles and sporadic events. See 'Climate cycles', *Green Murder* (Connor Court) by Prof Ian Plimer, p101ff, especially plot of atmospheric carbon dioxide and temperature, fig.4 p114

0.0012% 'FRIES THE PLANET'?

How many reasonable people would accept unquestioningly that a non-toxic gas making up 0.0012% of the atmosphere is a pollutant driving global warming? That's the amount of man-made CO^2 in our atmosphere. Australia's share, at 1.3% of that, is 0.0002%.[2]

These are uncontested facts – yet they are not generally known because they are never talked about in the public square – perhaps because they undermine the ruling orthodoxy that demonises fossil fuel emissions.

As geology professor Ian Plimer puts it, "Annual human emissions (3% of the total) of carbon dioxide are meant to drive global warming. This has never been shown. If it could be shown, then it would also have to be shown that natural emissions (97%) don't drive global warming."[3] The 3% of the total of 0.04% of carbon dioxide, 0.0012%, is what climate alarmists say is frying the planet. (That 97% from natural sources includes outgassing from the ocean, decomposing vegetation and other biomass, venting volcanoes, naturally occurring wildfires, and belches from ruminant animals.)

"There is no climate emergency. Climate science has degenerated into a discussion based on beliefs, not on sound self-critical science," according to CLINTEL's Climate Declaration[4], representing 1,122[5] scientists, founded in 2019 by emeritus professor of geophysics Guus Berkhout[6] and science journalist Marcel Crok.

---

2 https://www.worldometers.info/co2-emissions/australia-co2-emissions/ at time of writing shows Australia's share at 1.16%; the figure of 1.3% is in current general use

3 'Climate change delusion and the great Electricity ripoff' Emeritus Professor Ian Plimer (Connor Court).

4 https://clintel.org/world-climate-declaration/ is an independent foundation operating in the field of climate policy

5 As at August 7, 2022

6 Professor Guus Berkhout is a member of the Royal Netherlands Academy of Arts and Sciences (KNAW), The Netherlands Academy of Engineering (AcTI), honorary member of the Society of Exploration Geophysicists (SEG) as well as honorary member of the European Association of Geoscientists and Engineers (EAGE).

Scientists who defy climate alarmism have been effectively muzzled or ignored in the rush to condemn carbon dioxide as a dangerous pollutant.

Mass media failed to scrutinise the alarmist claims.

Policy makers around the world failed to fact check the alarmists; they accepted the groupthink.

That is why there is now a destructive energy crisis. Desperate times call for desperate measures; on July 7, 2022, The European Parliament backed EU rules labelling investments in nuclear power plants - and gas, a fossil fuel - as climate-friendly. Could coal be next to get the green light? Suddenly, European nations are panicking to try to rebuild their energy infrastructure. But since they've officially blocked most funding for non-green energy projects, the only way to get funds to rebuild fossil fuel infrastructure is to declare fossil fuels to be "green."

As an experiment, on June 20, 2022, when the Australian energy crisis deepened, I issued a media release through Medianet to 357 media outlets around Australia, and also distributed it to the Prime Minister and some 20 other politicians, mostly with relevant portfolios.

The press release was titled ENERGY CRISIS – A FAIILURE TO FACT CHECK. It included the text and the graph in this chapter. No politician responded beyond the automated acknowledgment of receipt. Only 26% of media recipients opened it. No media responded or published any of it – with the sole exception of The National Tribune – see *https://www.nationaltribune.com.au/energy-crisis-failure-to-fact-check/*

That astonishing lack of curiosity about the most consequential topic of our time has characterised media coverage of global warming and climate change topics for 33 years. The irony is that factually correct information is readily available, and has been for years - on the internet. Perhaps the usually reliable sources (such as learning and scientific institutions) were not providing it, having succumbed to the activist hysteria.

In this book, scientists in climate related fields provide the facts to enable policy makers to formulate policies based on evidence, not the group-thinkers shouting "WOLF!"

Their research should reassure children and inform adults: don't worry, be happy.... there is no climate emergency[7]. Man-made emissions are not a threat to the planet. There is no wolf there ...

The trace of man-made carbon dioxide in the atmosphere is about 0.0002%[8]; the elements in the atmosphere are nitrogen – 78.08% and oxygen – 20.95% = 99.03%; the rest is made up of argon, xenon, neon, hydrogen, helium, krypton ... and carbon dioxide, 0.04%.[9]

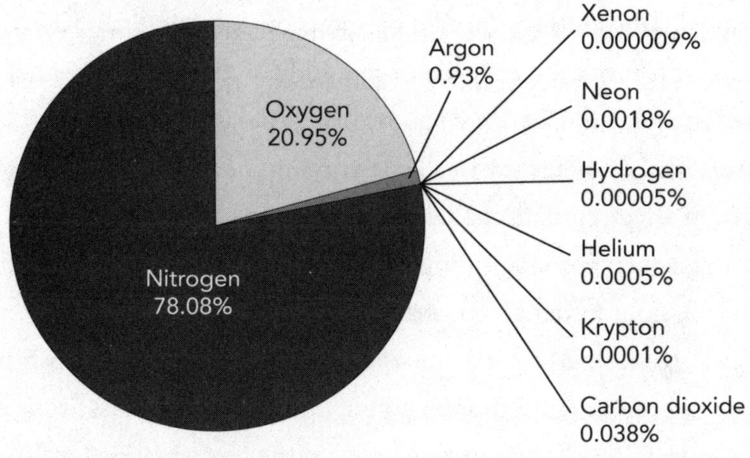

the composition of the Earth's atmosphere[10]

Australia's fossil fuel emissions account for 1.3% or so of that 0.0012%; that is 0.0002%.

Even if emissions really were the driver of warming, Australia's grasslands and forests annually suck up some 940 million tonnes of carbon dioxide – roughly double our emissions of 400-500 million tonnes.[11]

7 https://clintel.org/world-climate-declaration/

8 Being 3% of the total of 0.04% carbon dioxide in the atmosphere; 'Climate change delusion and the great Electricity ripoff' Emeritus Professor Ian Plimer (Connor Court).

9 https://en.wikipedia.org/wiki/Atmosphere_of_Earth

10 https://commons.wikimedia.org/wiki/File:Atmosphere3.svg

11 Emeritus Professor Ian Plimer, Spectator Australia, *Flat White*, May 24, 2022

These scientifically uncontested facts are never quoted in the public square, yet they are the most crucial pieces of information when considering the subject.

"In punching away at the clear shortcomings of the narrative of climate alarm, we have, perhaps, missed the most serious shortcoming: namely, that the whole narrative is pretty absurd. Of course, many people (though by no means all) have great difficulty entertaining this possibility. They can't believe that something so absurd could gain such universal acceptance." – Dr. Richard Lindzen, MIT Professor of Atmospheric Science and past UN IPCC contributor.

This book sets out the informed opinions of recognised climate scientists who have been engaged in the study of the complex field of climate. Included is a warning about current policy settings, that it is not a climate change emergency that is destroying the world,[12] but policies intended to affect 'climate change'.[13]

In short, contrary to the ruling orthodoxy, scientists do not regard human emissions as dangerous drivers of warming.

In my 2018 book, *Murder by the Prosecution* (Wilkinson Publishing), I explore cases of wrongful murder convictions. Of the six cases covered[14], all were especially analogous to the story of the boy who cried wolf... but the wolf wasn't there, in that we argue that the accused (and later convicted) were not at the crime scene at the time of the murders (or deaths). The juries relied on the evidence presented by the prosecution – and in each case, that evidence was variously false, inaccurate, incomplete or mischaracterised. The parallels with the absence of scientific 'evidence' about man-made global warming are inescapable.

---

12 "If green activism achieves its aims, the Third World will remain in poverty. Western countries will become impoverished and even more reliant on China which uses climate change as a weapon against the West." Green Murder (Connor Court), Emeritus Professor Ian Plimer; p 13

13 Where 'climate change' is in parentheses it refers to the commonly accepted notion that man-made emissions are responsible for global warming. Where climate change appears without parentheses it refers to historic and naturally occurring variations in climate.

14 Henry Keogh, Sue Neill-Fraser, Garry Nye, David Szach, Gordon Wood, Robert Xie; Neill-Fraser and Xie cases are at time of writing waiting for hearings in the High Court

'The wolf' has been expected many times in the climate change scenario, as dire predictions that the end is nigh tumble over each other, from then climate czar Tim Flannery (eg in 2007 he predicted: "So even the rain that falls isn't actually going to fill our dams and our river systems...") to organisations.

The Australian Bureau of Meteorology published a map on January 30, 2020, showing little chance of heavy rain in Australia in the period February to April 2020. A week later, the area with the lowest chance of heavy rain (Pilbara, WA) was hit by a drenching cyclone and eastern Australia was enjoying the normal conditions of drought, followed by flooding rains.

On July 3, 2012, *The Australian* reported a prediction from Catherine Pickering of Griffith University, headlined "Enjoy snow now – by 2020 it will be gone." Australia had record falls in 2019/20/21 and the snow seasons opened early – as it did again in 2022.

At the COP15 Climate Conference in December 2009, Al Gore predicted that the North Pole ice cap would be completely gone, ice free, by 2014.[15]

In 2009, the MSNBC documentary Future Earth 2025 claimed that: "As water levels drop, by 2017 Hoover Dam will no longer provide drinking water to Las Vegas, Tucson and San Diego. And it stops generating electricity to Los Angeles. And if nothing is done, the reservoir will be a dry hole by 2021."

In 2003, the VP for Field Programs at Defenders of Wildlife, Nina Fascione, made a terrifying forecast: "Frankly, it looks like we're on a crash course towards massive species extinctions in the next 20 years. We could lose 1/5 or 20% of our species within the next two decades. That's a very short amount of time."[16]

So many wolf alerts, yet no wolf sighted.

The lack of a wolf was no surprise to those who had read forecasting

15 Wattsupwiththat, December 16, 2018
16 Green Murder by Ian Plimer, page 173 from www.defenders.org April 18, 2003

scientists Kesten Green and Scott Armstrong's 2007 article "Global Warming: Forecasts by Scientists Versus Scientific Forecasts," in which they concluded:

"The forecasts in the [IPCC WG1] Report were not the outcome of scientific procedures. In effect, they were the opinions of scientists transformed by mathematics and obscured by complex writing. Research on forecasting has shown that experts' predictions are not useful in situations involving uncertainly and complexity. We have been unable to identify any scientific forecasts of global warming. Claims that the Earth will get warmer have no more credence than saying that it will get colder."

In a subsequent (2009) paper with physicist Willie Soon, "Validity of climate change forecasting for public policy decision making," they concluded that alarming warming trend forecasts had been beaten in the past by a simple benchmark forecast of no trend or no change, and would continue to be beaten by that forecast.

On the basis of the Green, Armstrong and Soon no-change forecast, Scott Armstrong challenged Al Gore to bet against that forecast with his "tipping point" warming alarm forecast of 2007. He declined, but Kesten Green has continued to monitor the "bet" as if it occurred using the IPCC's more moderate +3C/century "business as usual" projection of the time as a stand in for Gore's unquantified dangerous tipping point. Mr Gore would have lost the original 10 year bet had he taken it, and alarming warming continues to lose, as described at theclimatebet.com.

AUTHOR'S INTRODUCTION

Andrew L. Urban

I've been a journalist all my (long) working life; I am not an academic, not a scientist working in climate change, not a political operative nor a lobbyist. I have no hidden agenda or monetary interest in climate related energy policies. No shares in coal mines….no shares in renewables.

Researching and writing on this subject since 2011 (not counting my 2006 interview with Al Gore), I have observed a profound ignorance among citizens & their children, journalists, columnists, academics and policy makers. (When asked in 2019, Labor front bencher Tanya Plibersek confessed she doesn't know the amount of carbon dioxide in the atmosphere[17]. If you don't know, how can you say it's too much…? None of my family and friends knew the answer, either, until I told them.) That ignorance is evident on a global scale, and has led to malformed, needlessly destructive policies in most western countries.

But it is seeing children on the TV news crying in fear of the end of the world thanks to terrifying messaging and a lack of reliable information about global warming that has especially prompted compiling this report.

There is plenty of evidence about historical changes in the planet's climate, going back over 4 billion years – somewhat prior to the presence of man-made emissions. Propaganda has convinced many that carbon dioxide, a life source, is a dangerous pollutant in the atmosphere. All 0.04% of it. It has never been shown that this drives global warming. Claiming that 'the science' tells us to reduce fossil fuel emissions fails the test of truth. It is either dishonest or ignorant.

---

17 Daily Mail Australia, June 18, 2019

ABOUT THE AUTHOR

As a freelance journalist, Andrew has contributed (over 2,000 articles) to a wide range of publications, including *The Australian* (including *The Weekend Australian* & *The Australian Colour Magazine*), *The Sydney Morning Herald* (including *Good Weekend*), *Sun Herald* (Sydney), *Herald Sun* (Auckland), *The Spectator Australia* and previously to *The Bulletin*, *Cinema Papers*, the Qantas in-flight magazine, international publications including *The Japan Times* and others.

In 1986, he was the first accredited Australian journalist to file stories from the Cannes film festival, and has covered 20 Cannes festivals. He was Channel Host on the subscription TV channel, World Movies, for almost 5 years and since 2005, has presented Movies Now, a regular contemporary film appreciation course for the Centre for Continuing Education, University of Sydney.

Wearing a different hat, Andrew was the creator of and interviewer on Front Up, the 1990s prime time television program which aired on SBS for nine years, in which he fronted up to people in the streets of Australia for an impromptu conversation about their life experiences. The experience taught him to never assume anything about anybody.

Andrew was publisher and editor of Australia's online movie magazine, urbancinefile.com.au which he launched with his wife & business partner, Louise, in 1997, which they published for 20 years.

Since August 2011, Andrew has published nearly 100 articles on global warming/climate change issues on his blog, pursuedemocracy.org which provides over a decade of research helping to inform this book. Part of the first of these articles was taken from his submission to the Australian Select Senate Committee on the Scrutiny of New Taxes, August 15, 2011 (referring to a carbon tax).

Since 2013, Andrew has been investigating wrongful convictions for his blog, wrongfulconvictionsreport.org and his first book on the subject,

Murder by the Prosecution (Wilkinson Publishing, 2018). His other books (all Wilkinson Publishing) are *Margaret Cunneen SC – The Boxing Butterfly*; *Zelensky – the unlikely hero*; and his Ned Kelly Crime Fiction Award nominated first novel, *If You Promise Not To Tell.*

AWARDS

- The Royal Humane Society Testimonial (1961) for assisting in rescue of drowning man
- Front Up (SBS TV) - Mardi Gras 1997 Mainstream Media Award
- Urban Cinefile - Best Arts & Entertainment Site 1998, Telstra/ Financial Review Internet Awards

WARMING THE ROOM

How the global warming scenario was born in dishonesty

The anthropogenic global warming scenario was launched on June 23, 1988 in the US Senate committee with the testimony of James Hansen of NASA. (Hansen later compared coal freight trains to trains carrying victims to extermination camps.)

The scenario was born in dishonesty and has been characterised by it ever since. To emphasise the 'warming' at the congressional session, Hansen's Democrat ally, Senator Tim Wirth, scheduled the hearing on a day forecast to be the hottest in Washington that summer. In addition, Wirth sabotaged the air-conditioning the previous night, hoping to ensure the TV cameras could show everyone sweating in the heat. Wirth later told Deborah Amos (NPR News) how he did it: "What we did is that we went in the night before and opened all the windows, I will admit, right, so that the air conditioning wasn't working inside the room. And so when the hearing occurred, there was not only bliss, which is television cameras and double figures, but it was really hot … The wonderful Jim Hansen was wiping his brow at the table at the hearing, at the witness table, and giving this remarkable testimony."[18]

An example of the IPCC's methods was given by Dr. Richard Lindzen, IMT Professor Emeritus of Atmospheric Science and past UN IPCC contributor. He revealed that only a few scientists were involved in writing the IPCC 2001 Third Assessment Report. Although purported to speak for thousands of scientists, it was not thousands offering their consensus. Dr. Lindzen was a participant himself and said, "Each person who was an author wrote one or two pages in conjunction with someone

18 https://www.pbs.org/wgbh/pages/frontline/hotpolitics/etc/script.html

else. They travelled around the world several times a year for several years to write it, and the summary for policymakers had the input of about 13 of the scientists. Ultimately, it was written by representatives of governments, of environmental organizations like the Union of Concerned Scientists, and industrial organizations, each seeking their own benefit."

The unreliability of global warming enthusiasts was demonstrated when Al Gore's Oscar-winning documentary on global warming, An Inconvenient Truth, was criticised in October 2007 by a high court judge in Britain who highlighted what he said were "nine scientific errors" in the film.

The judge made his remarks when assessing a case brought by Stewart Dimmock, a Kent school governor and a member of a political group, the New party, who is opposed to a government plan to show the film in secondary schools.

The judge ruled that the film can still be shown in schools, as part of a climate change resources pack, but only if it is accompanied by fresh guidance notes to balance Gore's "one-sided" views. The "apocalyptic vision" presented in the film was not an impartial analysis of the science of climate change, he said.

The mistakes identified mainly deal with the predicted impacts of climate change, and include Gore's claims that a sea-level rise of up to 6m would be caused by melting in either west Antarctica or Greenland "in the near future".

The judge said: "This is distinctly alarmist and part of Mr Gore's 'wake-up call'." He accepted that melting of the ice would release this amount of water - "but only after, and over, millennia."

WHAT THE SCIENTISTS SAY

Man-made carbon dioxide is innocent

Writing in June 2022 on clintel.org[19], petrophysicist Andy May reported as follows: "Recently, the Biden administration has tried to use the powers of the SEC to force companies to disclose information on their supposed climate-related business risks through a proposed SEC rule. Two esteemed members of the CO_2 Coalition, Princeton Professor, emeritus, William Happer and MIT Professor, emeritus, Richard Lindzen have reviewed the proposed rule and filed a critical comment on the rule with the SEC. In addition, they have filed an amicus curiae court brief with the U.S. Court of Appeals for the Fifth Circuit stating that they do not believe there is a climate-related risk related to burning fossil fuels, and the resulting CO_2 and other greenhouse gas (GHG) emissions. This post discusses both filings made by Happer and Lindzen.

"It is utter nonsense"

The proposed SEC rule does not establish that there is a climate risk, they simply assume there is, based upon President Biden's Executive Order 13990 and the various IPCC Reports. The amicus curiae brief deals with the executive order and its underlying technical support document's (TDS) social cost of carbon (SCC) calculations.

Happer, who has studied possible CO_2 related climate change for over 40 years, succinctly states his opinion:

"There isn't a climate crisis. There will not be a climate crisis. It is utter nonsense."

19 https://clintel.org/critical-comments-by-happer-and-lindzen-on-sec-rule/

Lindzen has studied climate even longer than Happer, his comment: *"What historians will definitely wonder about in future centuries is how deeply flawed logic, obscured by shrewd and unrelenting propaganda, actually enabled a coalition of powerful special interests to convince nearly everyone in the world that carbon dioxide from human industry was a dangerous, planet-destroying toxin.*

"It will be remembered as the greatest mass delusion in the history of the world – that carbon dioxide, the life of plants, was considered for a time to be a deadly poison."

Happer and Lindzen point out that these proposed reporting requirements will cost U.S. businesses more than $6.4 billion and might interfere with their ability to raise the capital they need to expand production of oil, gas, and coal. This latter problem is expressly opposite of the SEC mandate from Congress. The rules will increase the cost of fossil fuels, provide no benefits, and expressly harm the poor, who pay a larger percentage of their income for fuel. Happer and Lindzen believe there are no climate risks related to burning fossil fuels; thus, estimating and reporting these imaginary costs will unnecessarily and unfairly restrict the companies' ability to borrow money because the additional risk reduces their value.

The SEC assumption is made because many prominent climate scientists share an opinion (the so-called "consensus") that humans are affecting the world's climate by burning fossil fuels and emitting large amounts of CO_2. They also share an opinion that this is a bad thing. However, in the scientific world, the opinions of scientists and politicians are not relevant. This is not to say that anthropogenic climate change or the possibility of an anthropogenic climate change disaster are disproven, it is just to say that no valid evidence exists to support these hypotheses.

Neither Happer nor Lindzen believe the SEC and TDS arguments are scientifically valid. They state that reliable scientific theories make predictions that are later validated by observations. They are not from a scientific consensus, government opinion, peer review, or manipulated data. In the words of Professor Richard Feynman, as quoted by Happer and Lindzen:

"[W]e compare the result of [a theory's] computation to nature, ... compare it directly with observations, to see if it works. If it disagrees with experiment, it is wrong. In that simple statement is the key to science." Richard Feynman, *The Character of Physical Law* (1965), p. 150.

Models have been created to show the hypothetical human-caused changes to climate and the supposed damage these changes might cause. Unfortunately, or fortunately, perhaps, the models do not compare well to observations. Using Feynman's rule, this invalidates the catastrophic climate change hypothesis. See here for more on the model/observations mismatch.

While the SEC can make rules mandating disclosure of valid risks to a business, it should not mandate disclosure of *imagined* risks that are not scientifically established.

To this point,

***

Back in 2010, 43 Fellows of the Royal Society (The United Kingdom's national academy of sciences, a Fellowship of some 1,600 of the world's most eminent scientists) wrote to its then president, Paul Nurse, to complain about the unscientific tone of the society's messages on climate change. Eight years later, a group of 33 current and former Fellows of the Geological Society wrote an open letter to their president in similar vein.

The letter notes that: "The IPCC position matches observations that almost half of the warming that has occurred over the last 150 or so years since industrialisation, had already happened by 1943, well before the rapid rise of industrial CO_2. This difference of opinion is critical, for if CO_2 did not cause the pre-1943 warming, the claimed consensus that Catastrophic AGW is caused by human CO_2 emissions since the Industrial Revolution, which is supported by GSL, must be mistaken.

"We also believe the GSL has a responsibility to refute the exaggerated claims that swirl around the fringes of the Climate Change debate, undermining the real science – such as that CO_2 and Climate Change cause:

more hurricanes, more rain, more drought, more asthma and now, even more terrorism (through drought in Africa), the exceptional cold and warm recorded over most of the sub-Arctic, Northern Hemisphere during the past winter and spring are what we should 'expect' from Global Warming.

"As this letter makes clear, it is not true that 97% of scientists unreservedly accept that AGW theory is fixed, or that carbon and CO_2 are 'pollutants' and their production should be penalised; how can the primary nutrient in photosynthesis be a pollutant? We also note that 700 scientists have made submissions to the US Senate expressing dissent from the consensus and 166 climate scientists issued a challenge to Ban Ki Moon on the eve of the Copenhagen Climate Summit in 2009 to provide proof of human induced global warming, which he did not do."

In Mark Steyn's *A Disgrace to the Profession* (Stockade Books, 2015), a comprehensive debunking of the consensus and in particular targeting Michael Mann's infamously influential but since discredited hockey stick graph, there are 100 scientists quoted who, far from supporting any consensus on global warming, raise their dissenting arguments and their concerns about misleading information.

"Some of them," Steyn explains, "are distinguished emeritus profs and Fellows of the Royal Society, but some are young up-and-comers. Many are from the heart of the Anglo-American climate establishment, but others are from Europe and elsewhere and don't quite understand why a small clique of outliers singlehandedly determines the "consensus" in the field …"

These scientists contradict the view that deniers are all right-wing nut jobs. Some are from the Scandinavian social-democrat left, some are hardcore Marxists. And Stephen McIntyre, the Toronto mining engineer who dismantled Mann's hockey stick graph (and is included in the book), is regarded as a 'conventional Trudeaupian liberal', as Steyn puts it. Mainstream scientists like Eduardo Zorita and Simon Tett are far from 'deniers'. This group of scientists demonstrate that there is no consensus to support CO_2 as the main driver of climate change. (Zorita is a Spanish paleoclimatologist; Tett is a climatologist working at the University of Edinburgh.)

On February 23, 2017, a petition was sent to then President Trump, signed by some 300 scientists, warning that the push to curtail carbon dioxide threatens to exacerbate poverty without improving the environment. In his accompanying letter to President Trump, MIT professor emeritus Richard Lindzen called on the United States and other nations to "change course on an outdated international agreement that targets minor greenhouse gases," starting with carbon dioxide.

"Since 2009, the US and other governments have undertaken actions with respect to global climate that are not scientifically justified and that already have, and will continue to cause, serious social and economic harm — with no environmental benefits," wrote Lindzen, a prominent atmospheric physicist. (until his retirement in 2013, Lindzen was Alfred P. Sloan Professor of Meteorology, Department of Earth, Atmospheric and Planetary Sciences at MIT)

Signatories of the attached petition include Dr David Evans, who was the Australian Greenhouse Office's carbon accounting modeler (1999 – 2005); the U.S. and international atmospheric scientists, meteorologists, physicists and professors, take issue with the United Nations Framework Convention on Climate Change [UNFCCC]. The 2016 Paris climate accord, which sets nonbinding emissions goals for nations, was drawn up under the auspices of the UNFCCC. Melbourne based climate data analyst John McLean, whose PhD thesis is titled "Prejudiced authors, prejudiced findings", claims that the climate data being used to propagate warming theories is essentially unreliable. He describes carbon dioxide emission mitigation policies as "a solution in search of a problem".

His lack of confidence in climate data grew out of the many weaknesses and failures of data since 1850 being used to underpin conclusions. For example:

- For the first 4 years the number of reporting observation stations in the Southern hemispheres was one - on the west coast of Indonesia;
- During the 1860s and 1870s Western Europe supplied a high proportion of Northern Hemisphere data despite being a very small proportion of the total area;
- About the same time a large proportion of Southern Hemisphere data was from the latitude bands 30S to 50S, this being much of the shipping route between Australia and Britain;
- In the 1850s Europe (and maybe the rest of the world) had hardly started to emerge from the Little Ice Age, so warming is no surprise;
- It's not until about 1950 that there is "decent coverage" of the two hemispheres, but it was only in the 1970s that sea surface temperature (SST) data was available for some parts of the Pacific.

McLean says these measurements were "often obtained by measuring the temperature of a sample of water drawn from 500mm below the surface and the technique was largely but not completely replaced by measuring the temperature of sea water that's drawn aboard to cool the engine, from a point two meters or more below the surface. In fact, about six different methods have been used, usually the data from all but one method somehow adjusted to match the single remaining method. It's all a bit of farce really. Argo buoys[20] seem to be more accurate but they've only been in common usage [in recent years]."

The Irish Climate Science Forum (ICFS) in cooperation with CLINTEL hosted a lecture by world-renowned now retired climate scientist Dr Richard Lindzen in April 2021. The online lecture was attended by around 200 people from around the world, including a group of climate activists who disturbed the talk.

Lindzen began his talk with the following observation:

"For about 33 years, many of us have been battling against climate hysteria. We have correctly noted:

The exaggerated sensitivity,

The role of other processes and natural internal variability,

The inconsistency with the paleoclimate record,

The absence of evidence for increased extremes, hurricanes, etc. and so on."

What he said next about the core status of climate discussions, is as important as his scientific analysis about CO_2:

"We have also pointed out the very real benefits of CO_2 and even of modest warming. And, as concerns government policies, we have been pretty ineffective. Indeed our efforts have done little other than to show (incorrectly) that we take the threat scenario seriously. In this talk, I want to make a tentative analysis of our failure.

"In punching away at the clear shortcomings of the narrative of climate alarm, we have, perhaps, missed the most serious shortcoming: namely, that the whole narrative is pretty absurd. Of course, many people (though

20 https://argo.ucsd.edu/about/

by no means all) have great difficulty entertaining this possibility. They can't believe that something so absurd could gain such universal acceptance. Consider the following situation. Your physician declares that your complete physical will consist in simply taking your temperature. This would immediately suggest something wrong with your physician. He further claims that if your temperature is 37.3C rather than between 36.1C and 37.2C you must be put on life support. Now you know he is certifiably insane. The same situation for climate (a comparably complex system with a much more poorly defined index, globally averaged temperature anomaly) is considered 'settled science.'

"In case you are wondering why this index is remarkably poor. I suspect that many people believe that there is an instrument that measures the Earth's temperature. As most of you know, that is not how the record was obtained.

"Obviously, the concept of an average surface temperature is meaningless. One can't very well average the Dead Sea with Mt. Everest. Instead, one takes 30 year annual or seasonal means at each station and averages the deviations from these averages. The results are referred to as annual or seasonal mean anomalies.

"...we are bombarded with claims that the impacts of this climate change include such things as obesity and the Syrian civil war. The claims of impacts are then circularly claimed to be overwhelming evidence of dangerous climate change. It doesn't matter that most of these claims are wrong and/or irrelevant. It doesn't matter that none of these claims can be related to CO_2 except via model projections. In almost all cases, even the model projections are non-existent. Somehow, the sheer volume of misinformation seems to overwhelm us. In case you retain any scepticism, there is John Kerry's claim that climate (unlike physics and chemistry) is simple enough for any child to understand. Presumably, if you can't see the existential danger of CO_2, you're a stupid denier.

"And, in case this situation isn't sufficiently bizarre, there is the governmental response. It is entirely analogous to a situation that a colleague, Bruce Everett, described. *After your physical, your physician tells you that you may have a fatal disease. He's not really sure, but he proposes a treatment that will be expensive and painful while offering no prospect of preventing the disease. When you ask why you would ever agree to such a thing, he says he just feels obligated to "do something".* That is precisely what the Paris Accord amounts to. However, the 'something' also gives governments the power to control the energy sector and this is something many governments cannot resist. "Information is unlikely to change this despite the fact that even the UN's IPCC acknowledges that their warming claims would only reduce the immensely expanded GDP by about 2-3% by the end of the century – something that is trivially manageable and hardly 'existential.'

"In trying to understand the success of this claim that climate change due to CO_2 is an existential threat, I propose to look at an analogous scare: the widespread fear in the US in the early 20th Century of an epidemic of feeblemindedness. I will also return to C.P. Snow's two-culture description in order to see why the alarmist scenario appeals primarily to the so-called educated elite rather than to the common people.

"Over twenty five years ago, I wrote a paper comparing the panic in the US in the early 1920s over an alleged epidemic of feeblemindedness with the current fear of cataclysmic climate change. ([1996] Science and politics: global warming and eugenics. in *Risks, Costs, and Lives Saved,* R. Hahn, editor, Oxford University Press, New York, 267pp [Chapter 5, 85-103])

"During this early period, the counterpart of Environmentalism was Eugenics. Instead of climate physics as the underlying science, we had genetics. And instead of overturning the energy economy, we had immigration restriction. Both advocacy movements were characteristically concerned with purity: environmentalism with the purity of the environment, eugenics with the purity of the gene pool. Interestingly, Eugenics did not start with a focus on genes. It was started around 1880 by biometricians who used statistical analysis to study human evolution.

"Among them were some of the founders of modern statistics like Pearson and Fisher. Given the mathematically sophisticated origin of the movement, it should come as no surprise that it didn't really catch on. It only became popular and fashionable when Mendelian genetics was rediscovered around 1900, and things like feeble mindedness were suggested to be associated with a single recessive gene. It is pretty clear that such movements need an easily understood, allegedly scientific but actually pretty absurd narrative. The people needing such narratives are not the ordinary citizen, but rather our educated elites.

"Prominent supporters of eugenics included Theodore Roosevelt, Margaret Sanger, the racist founder of Planned Parenthood, the Bishop of Ripon, George Bernard Shaw, Havelock Ellis, and many others. The supporters also included technically adept individuals who were not expert in genetics. Alexander Graham Bell for example. They also need a policy goal. In the early 1920s, Americans became concerned with immigration, and it was argued that America was threatened with an epidemic of feeblemindedness due allegedly to immigrants from Eastern and Southern Europe.

"Details of this situation are in my paper. The major takeaway points are the following:

Elites are always searching for ways to advertise their virtue and assert the authority they believe they are entitled to.

They view science as source of authority rather than a process, and they try to appropriate science, suitably and incorrectly simplified, as the basis for their movement.

Movements need goals, and these goals are generally embedded in legislation.

The effect of legislation long outlasts the alleged science. The Immigration Reduction Act of 1924 remained until 1964.

As long as scientists are rewarded for doing so, they are unlikely to oppose the exploitation of science."

When the world doesn't end, change the date

The scientific method delivers prosperity, yet scientific practice has become subject to corrupting influences from within and without the scientific community, say the authors of *The Scientific Method,* a blueprint for proper scientific methods, a high level study guide. In the process, they illustrate some of the pitfalls and weaknesses inherent in the way science is used … or abused. Below are just a handful of pertinent research-based observations that readers will find relevant to this book:

Failure to test plausible alternative hypotheses or, to use the term used by Armstrong and Green "advocacy research". Alternative hypotheses on climate include, e.g., any human contribution to global climate change is dwarfed by natural changes; it is not possible to predict global mean temperatures over decades and centuries better than no-trend; the net benefit of 3C of warming from late-18th Century temperatures would be greater than the net cost.

For an example of the persistence of an extreme belief, researchers studied a cult that predicted the "end of world" on a particular date. When that did not happen, the cult members increased their belief that the world would end, and **set a new date** (Festinger et al., 1956).

…A paper that challenges the current orthodoxy in a field might be **ignored regardless of its contribution** to scientific knowledge. (*eg papers of the late Bob Carter, Prof. Ian Plimer, Dr Jennifer Mahorasy, Dr Peter Ridd)*

People tend to avoid information that is inconsistent with their beliefs. If they cannot avoid new information that challenges their beliefs, they tend to evaluate it in ways that leave their prior beliefs unchanged. If they have no prior beliefs, they tend to adopt the beliefs of their friends, authorities, or groups with whom they identify. Experimental studies have shown that groups typically reject people with opinions that differ from the beliefs of people in the group. That happens even for newly formed groups when asked to decide on something about which they had no prior opinion. Pressure is directed against deviants, who face ostracism if they do not quickly agree with the group. (eg 'climate deniers'...also referenced in the section on 'Skepticism')

Confirmation bias is the tendency to interpret new evidence to support one's existing beliefs. It also involves efforts to seek confirming evidence and to avoid challenging information. If a challenge cannot be avoided, the source of the **challenging information is dismissed** as untrustworthy.

Governments can, and do, bestow authority on particular hypotheses. For example, Miller (2007) lists six "unassailable paradigms" that researchers cannot challenge without being silenced and **losing government grants.** He included "Cholesterol and saturated fats cause coronary artery disease," "Human activity is causing global warming," and "even a tiny amount of a toxin, such as radiation or cigarette will harm some people."

THE SCIENTIFIC METHOD: A Guide to Finding Useful Knowledge

J Scott Armstrong & Kesten Green, Cambridge University Press, 2022

J. Scott Armstrong is an author, forecasting and marketing expert, and a professor of Marketing at the Wharton School of the University of Pennsylvania.

Kesten Green is a leading researcher on forecasting methods and applications. He has developed new and better forecasting methods for business and public policy, and his research is widely cited. He has proposed and tested a unifying theory of forecasting, the Golden Rule of Forecasting, and is responsible for a review of evidence confirming the superiority of sophisticatedly simple forecasting methods over complex ones.

The authors had already published a paper (Green, Kesten C. and Armstrong, J. Scott, The Global Warming Alarm: Forecasts from the Structured Analogies Method (March 31, 2011.[21] which reveals the depth of betrayal of climate science.

" ... we searched the literature and asked experts to identify phenomena that were similar to the alarm currently being raised over dangerous manmade global warming. We obtained 71 proposed analogies. Of these, 26 met our criteria that the alarm be: (1) based on forecasts of human catastrophe arising from effects of human activity on the physical environment, (2) endorsed by experts, politicians and the media, and (3) that were accompanied by calls for strong action. None of the 26 alarms were based on scientific forecasting procedures. None of the alarming forecasts were accurate."

21 Available at SSRN: https://ssrn.com/abstract=1656056 or http://dx.doi.org/10.2139/ssrn.1656056)

CLIMATE CHANGE V CLIMATE VARIABILITY

Article 1 of the United Nations Framework Convention on Climate Change (UNFCCC) defines 'climate change' as:

"a change of climate which is attributed directly or indirectly to human activity that alters the composition of the global atmosphere and which is in addition to natural climate variability observed over comparable time periods."

The UNFCCC thus makes a distinction between climate *change* attributable to human activities altering the atmospheric composition, versus climate *variability* attributable to natural causes. This redefinition of 'climate change' to refer only to man-made climate change has effectively eliminated natural climate change from the public discussion on climate change. Any change that is observed over the past century, on whatever time scale, is implicitly assumed to be man-made. This assumption leads to connecting every unusual weather or climate event to man-made climate change from fossil fuel emissions.

The above observation was made by renowned climate scientist Dr Judith Curry[22]. One of the most important and useful essays on this subject was her article published in a Madrid newspaper coinciding with COP25, the December 2019 UN Climate Change Conference. Here is a selection of relevant points this respected scientist made:

- For the past three decades, the climate policy 'cart' has been way out in front of the scientific 'horse'. The 1992 Climate Change treaty

22 Dr Judith Curry is an American climatologist and former chair of the School of Earth and Atmospheric Sciences at the Georgia Institute of Technology. She was a member of the National Research Council's Climate Research Committee, published over a hundred scientific papers, and co-edited several major works. Curry retired from academia in 2017 at age 63 in frustration at the politicisation of climate science.

was signed by 190 countries *before* the balance of scientific evidence suggested even a *discernible* observed human influence on global climate. The 1997 Kyoto Protocol was implemented *before* we had any confidence that most of the recent warming was caused by humans. There has been tremendous political pressure on the scientists to present findings that would support these treaties, which has resulted in a drive to manufacture a scientific consensus on the dangers of manmade climate change.

- Fossil fuel emissions as the climate 'control knob' is a simple and seductive idea ... We have no idea how natural climate variability (solar, volcanoes, ocean circulations) will play out in the 21st century, and whether or not natural variability will dominate over manmade warming.
- We don't have a good understanding of how warming will influence extreme weather events. Land use and exploitation by humans is a far bigger issue than climate change for species extinction and ecosystem health. Local sea level rise has many causes and is dominated by sinking from land use in many of the most vulnerable locations.
- We have been told that the science of climate change is 'settled'. However, in climate science there has been a tension between the drive towards a scientific 'consensus' to support policy making, versus exploratory research that pushes forward the knowledge frontier. Climate science is characterized by a rapidly evolving knowledge base and disagreement among experts. Predictions of 21st century climate change are characterized by deep uncertainty.
- We have been told that climate change is an 'existential crisis.' However, based upon our current assessment of the science, the climate threat is not an existential one, even in its most alarming hypothetical incarnations. However, the perception of manmade

climate change as a near-term apocalypse and has narrowed the policy options that we're willing to consider.

- ... there is disagreement among experts regarding whether a rapid acceleration away from fossil fuels is the appropriate policy response. In any event, rapidly reducing emissions from fossil fuels and ameliorating the adverse impacts of extreme weather events in the near term increasingly looks like magical thinking.
- *The extreme rhetoric of the Extinction Rebellion and other activists is making political agreement on climate change policies more difficult. Exaggerating the dangers beyond credibility makes it difficult to take climate change seriously.
- The monomaniacal focus on elimination of fossil fuel emissions distracts our attention from the primary causes of many of our problems and effective solutions. Common sense strategies to reduce vulnerability to extreme weather events, improve environmental quality, develop better energy technologies, improve agricultural and land use practices, and better manage water resources can pave the way for a more prosperous and secure future. Each of these solutions is 'no regrets' – supporting climate change mitigation while improving human well-being. These strategies avoid the political gridlock surrounding the current policies and avoid costly policies that will have minimal near-term impacts on the climate. And finally, these strategies don't require agreement about the risks of uncontrolled greenhouse gas emissions."

***

"A global survey of 10,000 young people aged 16 to 25, including in Australia, Brazil, Finland, France, India, Nigeria, The Philippines,

Portugal, Britain and the US, has zeroed in on climate anxiety. The results, published in The Lancet Planetary Health (Dec. 2021), found 84 per cent of participants across all countries were worried about climate change, with 59 per cent very or extremely worried. More than half reported being sad, anxious, angry, powerless, helpless and guilty. More than 45 per cent of respondents said their feelings about climate change negatively affected their daily life and functioning, threequarters were fearful about the future and 83 per cent said they thought people had failed to take care of the planet. Climate anxiety and distress were correlated with perceived inadequate government response and associated feelings of betrayal," reported *The Australian's* environment editor, Graham Lloyd, on May 25, 2022.

That is hardly surprising. As Dr Curry points out, "the more attention a perceived danger gets, the more worried people become, leading to more news coverage and greater alarm. Because slowly and slightly increasing temperatures do not seem alarming, 'availability entrepreneurs' push extreme weather events, public health problems, human migration, etc. as being caused by man-made global warming – more of which is in store if we don't act NOW to reduce fossil fuel emissions.

"It is arguable that the constant and loud urgings of activists has gathered a whole range of other issues under the climate change banner, also without rational or evidentiary basis. 'Climate change' has become the receptacle for a myriad issues that are seen to derive from inaction on emissions. This unrealistic and volatile environment is further energised by the increasingly climate-focused political leadership around the western world."

THE IPCC

'The delinquent teenager mistaken for world's top climate expert'

The subtitle above is the subtitle of Canadian journalist Donna Lafromboise's book, *The Delinquent Teenager – who was mistaken for the world's top climate expert* (Connor Court). We know from the cover that the book is an 'IPCC expose', so we know who grows into a delinquent teenager. And no wonder, says Lafromboise, with over 100 godparents (all the countries represented on the Intergovernmental Panel on Climate Change) this child is bound to be spoilt.

"Having morphed into an obnoxious adolescent, the IPCC is now everyone's problem," she notes, and goes on to thoroughly disembowel the organisation, which produces the Climate Bible, the sequence of reports collated by the Panel and cited by Governments the world over. This Bible is the reason for carbon taxes, rising bills, costly regulations and "why everyone thinks carbon dioxide emissions are dangerous."

What Laframboise goes on to prove is that the Climate Bible is written, not by a meticulous upstanding professional in business attire, but "a slapdash, slovenly teenager who has trouble distinguishing right from wrong."

Her first indictment is that "the organisation is so arrogant, so used to being fawned over, that its leaders failed to take the most ordinary of precautions." It saw no need to discuss conflict of interest issues.

And while it has written down some of the rules of the road, it has never hired any traffic cops, she complains. The IPCC and much of the media claim that the reports it produces come from the "best talent available across the world." Laframboise cites several examples to show such claims to be unsupportable, whether in the field of extreme weather events, mosquitos or sea levels expertise.

Of several examples cited, one is that of a woman who was a research assistant at Australia's Monash University in 2008. After earning her PhD in 2009, she was hired by another university which boasted that she had already played a key role in both the 2001 and 2007 editions of the Climate Bible. The IPCC selected its 2001 authors in 1999; "this means its leadership decided she was a world class expert 10 years before she had earned her doctorate."

Another author didn't earn her PhD until 2010, yet in 1994 – 16 years earlier and three years before her first academic paper was published, she was one of just 21 people in the entire world selected to work on the first IPCC chapter that examined how climate change might affect human health.

A survey of IPCC policies and procedures by the InterAcademy Council, comprised of science bodies from around the world, found that even IPCC insiders were concerned that (in their own responses) "some of the lead authors are clearly not qualified..." and that there are far too many politically correct appointments. "The whole process is flawed by an excessive concern for geographical balance. All decisions are political before being scientific," echoing the most damning and frequent criticism made of the IPCC by its critics.

Laframboise finds that the IPCC has taken no steps to safeguard its reputation by maintaining a strict boundary between itself and green groups. The improper relationship is illustrated by the foreword to a 2007 Greenpeace publication written by none other than Rajendra Pachauri, then chairman of the IPCC (2002 - 2015). The following year he did it again, for another Greenpeace publication. And that's just for starters, says the author; the IPCC has filled many positions with Greenpeace or World Wildlife Fund activists. The 'activist scientist' is a species commonly found at the IPCC, as Laframboise shows.

The author devotes a chapter to climate modelling - and points out that the IPCC has not subjected climate models to rigorous evaluation. It simply asked climate modellers to evaluate their own handiwork. "This is like asking parents to rate their own children's attractiveness," she quips.

She says that if you are one of those who still believe that the IPCC carefully verifies the research on which its most important conclusions are based, you will be disappointed if not devastated to learn - from the mouths of those who have worked inside the beast - that it does a very poor job of that. Laframboise cites some of the answers from insiders published in that InterAcademy Council survey: "Quality assurance and error identification is not existent," says one. Another confirms that: "As far as I can tell, there is no data quality assurance associated with what the IPCC is doing." There are several others along these lines.

The peer review is just as flawed, contrary to Pachauri's response to critics; "The IPCC studies only peer reviewed science..." This notion (promulgated by all IPCC supporters, including former President Obama and his then science advisor John Holdren) has been disseminated by journalists and governments - and scientists, Laframboise says accusingly. But peer review - a notion as misused as it is at the IPCC - is harnessed by the IPCC both "as a shield behind which it hides, and a sword with which it skewers dissenting voices", as Laframboise puts it.

Her book forensically exposes instances of what can only be called scientific fraud, ethical failures and political massaging of information. Her accusations are all backed by dozens of citations.

Canada's self-described feminist and activist journalist, Laframboise concludes that the IPCC should be disbanded. "A new, untainted group of people must start from scratch. And this time, it should be ruled by genuine checks and balances, environmentalists should be kept well away for everyone's sake". As for the media it "should have subjected the IPCC to the same kind of scrutiny that keeps other organisations honest."

In the following extracts from a timeline compiled by Tony Thomas of *Quadrant Magazine*, the IPCC's reliability is under further scrutiny:

> **1988:** *Maurice Strong, executive director of the UN Environmental Program, helped get the IPCC set up as a combination of UNEP and the World Meteorological Organisation. Strong also organised the Rio Earth Summit of 1992, which woke up the West's politicians to a new, feel-virtuous campaign involving vast potential tax inflows. In 2007, while investigating corruption in the Iraq Food for Oil program, the FBI came across a cheque to Strong for US$998,000 by a corrupt South Korean businessman via a Jordanian bank. Strong hastened to the first plane for Beijing, China having no extradition formalities with the US, and lived there until his death in 2015 at 86, the cheque still unexplained.*
>
> **1999:** *To make its warming story stick, the IPCC needed to show that 20th-century warming was 'unprecedented', but the Medieval Warming Period (when Greenland grew grapes) was a fly in the IPCC ointment. A then-youthful scientist, Michael Mann, constructed a 1000-year temperature record using proxies such as tree rings. This graph, the infamously inaccurate 'hockey stick', erased both the Medieval Warming and the Little Ice Age (1550-1850) that followed.*
>
> *The IPCC featured the Mann chart seven times in its 2001 report, the graph becoming for a while the virtual logo and icon of the organisation. But Mann refused all requests to make his data and algorithms public for verification. In the IPCC's 2007 report, the hitherto famous 'hockey stick' was downgraded to one mention of its controversial nature. In 2017 Mann was still hiding his data, even in defiance of a Canadian court ruling last July that it be provided to defendants in a libel suit Mann himself had initiated. In Washington, where another libel suit against columnist Mark Steyn has been*

bogged down for years, he has been no less coy about revealing the secrets of his climate modelling.

2007: *The IPCC Fourth Report was published claiming warming would melt the Himalayan glaciers by 2035 and deprive billions on the sub-continent of fresh water. The claim was based on a chat between a supposed glacier expert, Syen Hasnain, and a magazine reporter. The relevant page of the IPCC report now contains no less than nine "erratas", even conceding dud arithmetic. When a genuine glacier scientist, Vijay Raina, challenged the 2035 claim, noting that melting would actually take many centuries, IPCC chair and dirty old man Rajendra Pachauri derided him for 'voodoo science'. Pachauri then appointed Syed Hasnain, the original source of the erroneous 2035 claim, to his think-tank and secured millions of dollars in grants to study the faked melting-glacier crisis.*

2010: *Unwilling to further endorse the IPCC's credibility, the InterAcademy Council (the executive composed of 11 national science bodies) ordered an audit of the IPCC. It urged Pachauri to quit but he refused. The audit found "significant shortcomings in each [i.e. every] major step of IPCC's assessment process".*

The audit results were swept under the rug by the Australian Academy of Science, notwithstanding that its then-president Kurt Lambeck, was a prominent member[23] *of the international audit.* (There is no suggestion that Lambeck was involved in the suppression.)

November 14, 2010: *Ottmar Edenhofer, then co-chair of IPCC Working Group III, is quoted by the Zuricher Zeitung*[24]*:*

"Climate policy has almost nothing to do anymore with environmental protection, says the German economist and IPCC

23 https://quadrant.org.au/magazine/2012/06/our-planet-saving-science-lobbyist-the-integrity-of-the-australian-academy-of-science/

24 https://wattsupwiththat.com/2010/11/18/ipcc-official-climate-policy-is-redistributing-the-worlds-wealth/

official Ottmar Edenhofer. The next world climate summit in Cancun is actually an economy summit during which the distribution of the world's resources will be negotiated."

(Kesten Green was a presenter at the Alpbach Forum in 2008, and when he challenged Ottmar (also a presenter) that "the predicted warming, should it occur, would have positive as well as negative externalities, he agreed, and did not have a proposed solution in "climate policy"… so I like to think that I pushed him to fessing up that the policies have nothing to do with the climate or effects of climate change, or changing climate.")

February 24, 2015: *IPCC chair (2002-2015) Pachauri, 74, quits abruptly on being charged by New Delhi police with multiple counts of sexual stalking and harassment involving a 29-year-old female employee. Pachauri denies guilt, initially claiming that hundreds of smutty texts to the fixation of his frustrated affections were written by a hacker who somehow took over his emails, text-messaging and WhattsApp accounts.*

2017: *Because the narrative of sharply rising global temperatures has failed, alarmist scientists have switched their narrative to "extreme weather" and its alleged hazards. This new story flies in the face of the IPCC's own SREX special report of 2012 on extreme weather, which conceded that warming could well reduce extremes, rather than increase them. Further, it would be 20-30 years before any climate effects on extreme weather would even be detectable against natural climate variability. The 2013 IPCC report broadly endorsed those findings.*

Thomas has also reported: "The IPCC's melting-glacier scandal of 2010 and the "Climategate" e-mail scandals (2009 and 2011) have arguably forced the IPCC into a more disciplined approach, with the determination not to be further caught out on scientific bias."

But it was too little, too late, Thomas noted. "The fruits of this new approach emerged in November 2011 with the IPCC's special draft report on extreme weather events. Thanks to anodyne IPCC press releases, the mass media ... failed to notice a bombshell finding. Translated from long-winded science-style language, it says:

1. for the next twenty to thirty years, man-made warming effects on climate extremes will be swamped by natural climate variability;
2. the man-made warming may even be beneficial by reducing the number of extreme events; and
3. neither IPCC models nor emissions forecasting are good enough to forecast extreme weather events up to the end of the century."

IPCC STORMS IN A TEACUP

In early July 2022, Australia's Prime Minister, Anthony Albanese, was commenting on the floods which had yet again devastated large regions north of Sydney. He told the media how climate change was causing ever increasing extreme weather events. Neither the media nor any of his advisers corrected him. Such false proclamations by political leaders feed into the general ignorance and misinformation about the subject. 'The science tells us…' has become a false claim that excuses plain ignorance and disrespects science.

In his letter of resignation from participating in the Fourth Assessment Report of the Intergovernmental Panel on Climate Change, Chris Landsea was aghast that his colleagues "would utilize the media to push an unsupported agenda that recent hurricane activity has been due to global warming."

In his letter, he pointed out "All previous and current research in the area of hurricane variability has shown no reliable, long-term trend up in the frequency or intensity of tropical cyclones, either in the Atlantic or any other basin. The IPCC assessments in 1995 and 2001 also concluded that there was no global warming signal found in the hurricane record. Moreover, the evidence is quite strong and supported by the most recent credible studies that any impact in the future from global warming upon hurricane will likely be quite small. The latest results from the Geophysical Fluid Dynamics Laboratory (Knutson and Tuleya, Journal of Climate, 2004) suggest that by around 2080, hurricanes may have winds and rainfall about 5% more intense than today. It has been proposed that even this tiny change may be an exaggeration as to what may happen by the end of the 21st Century (Michaels, Knappenberger, and Landsea, Journal of Climate, 2005, submitted).

***

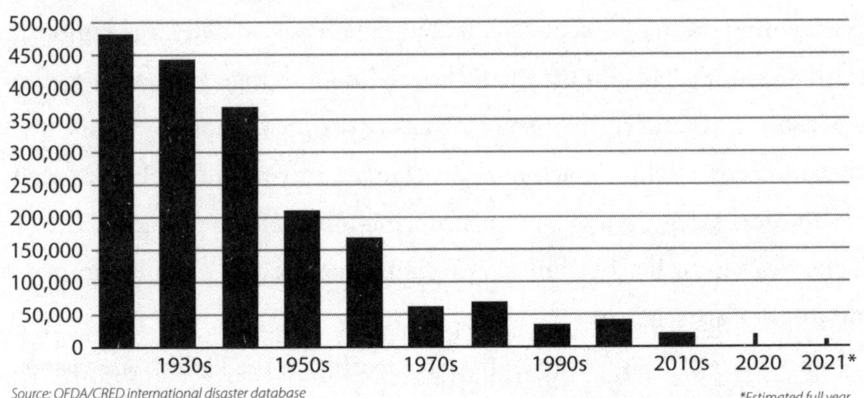

When the IPCC was set up over 30 years ago its objective was "to stabilize greenhouse gas concentrations in the atmosphere at a level that would prevent dangerous anthropogenic [i.e., human-induced] interference with the climate system". The original mandate from the UN Framework Convention on Climate Change (UNFCCC) for the IPCC was to address ***'dangerous human-caused climate change'***. That set the agenda, which became the ruling orthodoxy, a circular argument that **starts with the conclusion it is trying to prove.**

"Research and other professional activities are professionally rewarded only if they are channeled in certain directions approved by a politicized academic establishment — funding, ease of getting your papers published, getting hired in prestigious positions, appointments to prestigious committees and boards, professional recognition, etc. How young scientists are to navigate all this is beyond me, and it often becomes a battle of scientific integrity versus career suicide," Dr Curry wrote on her blog after taking early retirement on January 1, 2017 as Emeritus Professor of Earth and Atmospheric Sciences. She is also President of Climate Forecast Applications Network (CFAN).

Dr Curry points out that it is "an empirical fact that the Earth's climate has warmed overall for at least the past century. However, we do not know how much humans have contributed to this warming and there is disagreement among scientists as to whether human-caused emissions of greenhouse gases is the dominant cause of recent warming, relative to natural causes." Dr Curry made this statement on March 29, 2017 as part of her written evidence to the US House of Representatives Committee on Science, Space & Technology (Climate Science: Assumptions, Policy Implications and the Scientific Method).

The most famous examples of unscientific behaviour were the manipulated hockey stick temperature graph from Michael Mann's team (as mentioned earlier) and its propagation and embellishment by Al Gore's film, *The Inconvenient Truth* - and the so called Climategate email scandal, which revealed that Phil Jones and other climate scientists in his team were also trying to manipulate data to fit their agenda. *(The frequent practice of showing images of extreme weather events and attributing the events to 'climate change' is also unscientific and misleading; the picture of a burning house doesn't tell you how the fire started.)*

Professor John Christy, Distinguished Professor of Atmospheric Science, Alabama's State Climatologist and Director of the Earth System Science Centre at The University of Alabama in Huntsville, also gave evidence at the Science Committee hearings. Christy's evidence shows that observed warming has been significantly less than models have predicted. " … if one follows the scientific method … the average model trend fails to represent the actual trend of the past 38 years by a highly significant amount. As a result, applying the traditional scientific method, one would accept this failure and not promote the model trends as something truthful about the recent past or the future. Rather, the scientist would return to the project and seek to understand why the failure occurred. The most obvious answer is that the models are simply too sensitive to the extra GHGs [green house gases] that are being added to both the model and the real world." In other words, human made emissions of CO_2 have less impact than feared. Less impact than claimed, is perhaps the more accurate way of putting it.

HOLES IN THE RULING ORTHODOXY

Truth the casualty in climate change debates

One of the most dishonest entries that damaged public knowledge in a long list of mischaracterisations of climate change related 'science' is the claim that "97% of scientists" agree that global warming is driven by man-made emissions. These claims are based on a paper published by a team led by Australia's John Cook[25] in the journal Environmental Research Letters. In a leak from an internal forum at the global warming activist website Skeptical Science, Cook explained the paper's purpose was to establish the existence of a consensus – even of it had to be manufactured.

"It's essential that the public understands that there's a scientific consensus on AGW. So (Skeptical Science activists) Jim Powell, Dana (Nuccitelli) and I have been working on something over the last few months that we hope will have a game-changing impact on the public perception of consensus. Basically, we hope to establish that not only is there a consensus, there is a strengthening consensus." Ian Plimer calls it the zombie statistic because it won't die. Of the 11,944 papers studied, only 41 (0.3%) authors explicitly stated that they thought humans caused most of the warming. "In the physical sciences we would call Cook's work fraud. In the social sciences it is career advancement," he quips.

US research scholar Michelle Stirling[26] writes: "A claimed consensus is a powerful tool for driving policy but an inappropriate and unethical

25 John Cook is a Research Assistant Professor at the Centre for Climate Change Communication, George Mason University, Virginia. He holds a PhD in cognitive psychology at the University of Western Australia and a Bachelor of Science at the University of Queensland.

26 Michelle Stirling works for Friends of Science, which receives funding from the fossil fuel industries.

means of conducting scientific inquiry or informing the public. This fakery from zealots has been collectively adopted as driver of policy by the entire Western political class (the Chinese aren't that stupid). And we will pay for it in more ways than one. We have already got the electricity bill."

In her editor's introduction to the book, *Climate Change: The Facts 2020* (Institute of Public Affairs), Jennifer Marohasy points out, "It could be, as Ansley Kellow explains in Chapter 18, that climate science has demonstrated noble cause corruption, where the ends justify the means or where there is a belief in a moral commitment in producing evidence - any evidence - intended to have an influence on the political landscape. A fact is something that is known or proven to be true. A scientific fact cannot be established by a consensus of opinion or by the popular vote or because it is morally good. The fact may contain defensive information but may still be true."

Many scientists have been deeply disturbed by the manipulation of data and similar falsifications to hold up a global warming theory. That birthmark, *dishonesty,* has never been erased.

Among the 100 honest scientists who do not subscribe to dishonest methods, are these who are quoted in Mark Steyn's book *A disgrace to the profession* (Stockade Books):

PROFESSOR DARREL INCE, PHD Professor of Computing at the Open University's Centre for Research in Computing in the United Kingdom. Author of peer-reviewed papers published by Empirical Software Engineering and other journals.

DR HENDRIK TENNEKES, PHD Former Director of Research at the Royal Dutch Meteorological Institute and member of the Royal Netherlands Academy of Arts and Science. Former Professor of Aeronautical Engineering at Pennsylvania State University, author of The Simple Science Of Flight From Insects To Jumbo Jets (MIT Press,

1997) and co-author of the classic A First Course In Turbulence (MIT Press, 1972).

DR MICHAEL R FOX, PHD (1936-2011) Nuclear scientist, Professor of Chemistry at Idaho State University and researcher at the National Engineering Laboratory. Chairman of the American Nuclear Society's Public Information Committee.

DR HAMISH CAMPBELL, PHD Geologist and paleontologist with New Zealand's Institute of Geological and Nuclear Sciences (GNS Science), and scientist in residence at Te Papa Tongarewa, the national museum of New Zealand. Former President of the New Zealand Association of Scientists, Companion of the Royal Society of New Zealand, and Member of the Geological Society of New Zealand.

PROFESSOR IVAR GIAEVER, PHD Winner of the 1973 Nobel Prize in Physics, with Leo Esaki and Brian Josephson, "for their discoveries regarding tunnelling phenomena in solids". Professor-at-large at the University of Oslo, and Professor Emeritus at the Rensselaer Polytechnic Institute. Recipient of the Oliver E Buckley Condensed Matter Prize from the American Physical Society and the Zworykin Award from the National Academy of Engineering. Member of the Norwegian Academy of Science and Letters.

And of course, Australia's acclaimed geologist Ian Plimer, has written and spoken extensively about the absence of evidence proving man made CO_2 to be the climate culprit. His extensively researched 2021 book (with 1,667 citations), *Green Murder* (Connor Court) is a source for several entries and footnotes in this book.

Professor William Happer, Cyrus Fogg Brackett Professor of Physics at Princeton says: "The existence of climate variability in the past has long been an embarrassment to those who claim that all climate change is due to man and that man can control it ... I was very surprised when I first saw the celebrated "hockey stick curve," in the Third Assessment Report of the IPCC. I could hardly believe my eyes. Both the Little Ice Age and the Medieval Warm Period were gone, and the newly revised temperature of the world since the year 1000 had suddenly become absolutely flat until the last hundred years when it shot up like the blade on a hockey stick…"

The definition of climate change deniers as summarised by Wikipedia:

> *Climate change denial, or global warming denial, is denial, dismissal, or* ***unwarranted doubt*** *that contradicts the scientific consensus on climate change, including the extent to which it is caused by humans, its effects on nature and human society, or the potential of adaptation to global warming by human actions.*

Unwarranted doubt? Wikipedia reveals its bias; there goes its credibility and its authority. This bias is evident throughout the pages of Wikipedia. Another example:

> *The Nongovernmental International Panel on Climate Change is a* ***climate change denial advocacy*** *organisation set up by S. Fred Singer's Science & Environmental Policy Project, and later supported by the Heartland Institute lobbying group, in opposition to the assessment reports of the Intergovernmental Panel on Climate Change on the issue of global warming…*

We certainly can't accuse Wikipedia of being a disinterested repository of facts.

SCIENTISTS, CREDIBILITY, BIAS

"The most important fact about climate science, often overlooked, is that scientists disagree about the environmental impacts of the combustion of fossil fuels on the global climate. There is no survey or study showing "consensus" on the most important scientific issues, despite frequent claims by advocates to the contrary," write Craig Idso, S. Fred Singer and Robert M. Carter (1942-2016) in their book *Why scientists disagree about global warming* (Heartland).

They go on to explain that "Scientists disagree about the causes and consequences of climate for several reasons. Climate is an interdisciplinary subject requiring insights from many fields. Very few scholars have mastery of more than one or two of these disciplines. Fundamental uncertainties arise from insufficient observational evidence, disagreements over how to interpret data, and how to set the parameters of models. The Intergovernmental Panel on Climate Change (IPCC), created to find and disseminate research finding a human impact on global climate, is not a credible source. It is agenda-driven, a political rather than scientific body, and some allege it is corrupt. Finally, climate scientists, like all humans, can be biased. Origins of bias include careerism, grant-seeking, political views, and confirmation bias.

"Probably the only "consensus" among climate scientists is that human activities can have an effect on local climate and that the sum of such local effects could hypothetically rise to the level of an observable global signal. The key questions to be answered, however, are whether the human global signal is large enough to be measured and if it is, does it represent, or is it likely to become, a dangerous change outside the range of natural variability? On these questions, an energetic scientific debate is taking place on the pages of peer-reviewed science journals.

"Rather than rely exclusively on IPCC for scientific advice, policymakers should seek out advice from independent, nongovernment organizations and scientists who are free of financial and political conflicts of interest. NIPCC's (Nongovernmental International Panel on Climate Change) conclusion, drawn from its extensive review of the scientific evidence, is that any human global climate impact is within the background variability of the natural climate system and is not dangerous.

"In the face of such facts, the most prudent climate policy is to prepare for and adapt to extreme climate events and changes regardless of their origin. Adaptive planning for future hazardous climate events and change should be tailored to provide responses to the known rates, magnitudes, and risks of natural change."

THE CHIEF SCIENTIST & CLIMATE ADVICE

We heard from State and Federal Chief Medical Officers on a daily basis during the heaviest phase of the Covid epidemic. They made the rules. They were The Authority Who Must Be Obeyed. Yet we never once heard from the Chief Scientist during the many and heavy phases of climate and energy policy formulation over the years. That silence was always taken to signal agreement with Government policies in matters based on science.

Well, that in fact has been true, unfortunately, as we'll see shortly. Instead of scrutiny, the public got complicity from scientists who knew better. They risked their careers unless they agreed with the ruling orthodoxy.

Perhaps the reason the Chief Scientists was not at the podium for climate and energy announcements was that such policy is always based on assumptions, propaganda and the received wisdom of the IPCC – exaggerated to fit the orthodoxy. By exaggerated, we probably mean a pattern of dishonesty, from the deliberately warmed room for its 'birth' to the infamous hockey stick graph to the Climategate scandal and much besides. Fear was the driver of the Doom Train that was and is always just over the horizon.

It was left to the IPCC to brief the world's naïve politicians on the dangers of 'climate change', as it has been labelled for ease of reference. "Do you believe in climate change?" is the qualifier for jobs as well as social acceptance. They mean 'do you believe that man-made emissions drive global warming'?

The average person is not expected to know the composition of the atmosphere. The promoters of alarmism rely on thst ignorance.

Those who formulate policy have a responsibility to check the facts on which they base decisions.

According to the website, the Chief Scientist's mission is to:

- Provide authoritative, independent science advice
- Champion Australia's science and research system
- Improve Australia's scientific capability

In the section Advice to Government, the website lists a number of areas. First on the list is its role in The National Science and Technology Council, which is "responsible for providing advice to the Prime Minister and other Ministers on important science and technology issues facing Australia." Surely that embraces the fraught issues around climate science? Australia's Chief Scientist, Dr Cathy Foley, is the Executive Officer of the Council.

The appointed scientific expert members of the Council are:

- Professor Genevieve Bell, AO,
- Professor Debra Henly,
- Professor Brian Schmidt, AC,
- Professor Fiona Wood, AM,
- Emeritus Professor Cheryl Praeger AC,
- Associate Professor Jeremy Brownlie

"The Council is the preeminent forum for providing scientific and technological advice for government policy and priorities."

For example:

> *"Rapid Response Information (RRI) Reports are prepared on behalf of the National Science and Technology Council (the Council) to deliver timely responses on specific questions raised by the Australian Government.*
>
> *Introduced as part of the Council's refreshed Terms of Reference,*

> *the reports provide a more responsive mechanism for Ministers to access science and technology advice in a consistent and policy relevant manner.*
>
> *"Latest Rapid Response reports:*
>
> *• Space Industry and the STEM workforce*
>
> *What are the growth areas in domestic STEM skills to support jobs in the space industry, and how can these be addressed by the tertiary (university and relevant VET) sector?"*

There was no mention of a Rapid Response report on what lessening of global warming would result in the policy-forced forced reduction of Australia's emissions of the magnitude of 40 -50%. Australia's emissions are a tad over 1% of the world's total. As this book reports elsewhere, Australia's grasslands and forests annually suck up some 940 million tonnes of carbon dioxide – roughly double our total emissions of 400-500 million tonnes. So even though emissions matter to climate activists, they are unaware of (or would rather not know) relevant facts.

Admirably, the Chief Scientist's work involves Horizon Scanning: "reports by the Australian Council of Learned Academies (ACOLA) draw on the deep disciplinary expertise from within Australia's Learned Academies to analyse the future, navigate change and highlight opportunities for the nation. As interdisciplinary studies, ACOLA's reports include economic, social, cultural and environmental perspectives to provide well-considered findings that inform complete policy responses to significant scientific and technological change."

The latest report, in November 2020, was ***The Internet of Things: Maximising the benefit of deployment in Australia.***

But still nothing on climate science…

On July 28, 2022, I wrote to the Chief Scientist's office to ask as follows (that freelance article I mentioned has turned into this chapter):

Dr Cathy Foley,

In preparation for an article for my freelance clients, I have been exploring your corporate website. I have found a number of submissions under Advice to Government, but perhaps this section is not fully populated with all the advice given by you or your predecessor, Dr Alan Finkel. I could not find any advice as to the scientific rationale for the policies being implemented for the urgent reduction of fossil fuel emissions. As this is perhaps the most consequential policy subject of our time, I would be grateful for access to any such analysis.

Would you please respond to the following questions:

1 *Given that* CO^2 *comprises just 0.04% of our atmosphere, are you able to provide copies of reports (or links to) that set out the empirical basis for ascribing global warming to carbon dioxide? Further, are you likewise able to provide copies of reports (or links to) that demonstrate how 3% of that, the man-made element, drives warming but the 97% naturally occurring does not?*

2 *Under your duty to 'provide authoritative, independent science advice' to Government, please advise:*

   A) Have any Governments sought your advice or have you offered advice on the level of carbon dioxide in our atmosphere and its origins?

   B) If so, when and what, in summary, was the request and the advice?

   C) If not, would you welcome such a request?

I am asking these questions because the Government has not provided answers by way of policy rationale - and there are statements by your predecessor for which I could find no scientific rationale. For example:

Dr Alan Finkle *(emphasis added)*

3/11/2016 Preliminary Report of the National Electricity Market (NEM) Security Review.

"But human progress has come at a price, ***in the form of climate change.*** *And, with good will and time on our side, electricity generation can be decarbonised. Whether it's hydroelectric dams, or solar and wind, there are many viable technology paths to cutting emissions."*

9/6/2017 Final Report of the *Independent Review into the Future Security of the National Electricity Market*

"The blueprint released today presents the essential elements for a strategic plan for our electricity future. It is up to Federal, State and Territory Governments to take these recommendations, make decisions, add detail and drive it forward."

The blueprint will deliver four key benefits for the electricity system:

- *future reliability*
- *increased security*
- *rewarding consumers*
- ***lower emissions***

The report uses three pillars to achieve these outcomes: orderly transition measures, system planning and stronger governance.

Under the orderly transition pillar, the Review Panel concluded that a ***Clean Energy Target*** *is the most effective mechanism* ***to reduce emissions*** *while supporting security and reliability.*

Existing large electricity generators will be required to give a ***three years' notice of closure****. This will signal investment opportunities for new generation and give communities time* ***to adjust to the loss of a large employer****.*

10/7/2019 A message from the Chief Scientist: Australia's Hydrogen Potential

*"**Imagine a zero-emissions fuel** that exists on Earth in abundance, can be easily extracted using basic chemistry and offers jobs and investment in Australia for decades to come.*

That substance exists: it's called hydrogen. Just like natural gas, hydrogen can be used for heating and cooking in homes. Instead of petrol and diesel, hydrogen fuel cells can power electric trucks, trains and cars.

We've produced it in large volumes for more than a century, for use as a feedstock in industry; and it's been shipped and stored with an exemplary safety record for all that time. Now hydrogen is surging to the top of the global decarbonisation agenda."

Dr Finkel's statements demonstrate his acceptance of the ruling orthodoxy on the matter of climate science: fossil fuel emissions drive warming. Dr Finkel did, however, admit to the Senate that eliminating ALL of Australia's emissions of 1.3% of the world's output, would do "virtually nothing" for the climate. (In the scenario that emissions drive warming.)

What does the current Chief Scientist have to say in response to my questions?

August 11, 2022:

"From a spokesperson for Australia's Chief Scientist, Dr Cathy Foley: Dr Foley has confidence in the scientific process and in what the experts in this field are telling us about the changing climate. Dr Foley has not provided specific advice to the Government on this matter."

During the 'Covid emergency', every Chief Health Officer in the country gave daily health advice to governments, yet in what alarmists are calling a 'climate emergency', the nation's Chief Scientist has not provided any climate advice to Government.

I also received a supplementary email:

"You will be aware, of course, that Dr Foley is a member of the Climate Change Authority. The authority has just released its Review of International Offsets."

The review begins:

11 August 2022: The Climate Change Authority's Review of International Offsets finds the international carbon market is still evolving in response to the Paris Agreement and calls for publication of a National Carbon Market Strategy that makes the most of this opportunity for Australia to accelerate ambition on emissions reduction.

The review finds that while carbon is priced and traded in Australia, the market is fragmented, inefficient and complicated.

IF IT'S SCIENCE, WHY THE POLITICS?

The fervent climate activist is like a dog with a bone about 'climate change' – not the scientific fact of it but the socio-political message of the term. Instead of debating the subject, their first tendency is aggressive denigration; name calling, the last resort. It's as if positive information – that human activity is not endangering the planet – was unwelcome; fear of catastrophe is preferred. (That might explain why children's charities offered the royalties from this book refused to accept them. See DONATION CENSORED)

Why not engage? Why not talk about the evidence on which those claims rely? Why not cite the evidence for how much carbon dioxide in the atmosphere produces 1C of warming, say? (Is it perhaps because if there was such evidence, we wouldn't be having this argument?)

Why is it the Labour/Greens who most noisily advocate for 'action on climate change'. That poorly framed demand has infected several members of the Liberal Party, too, and like all converts, are even more enthusiastic; eg the Victorian Liberals who overtook Labour's 43% emission reduction target with 50%. Cop that! (COP27, Egypt, November 2022)

The most ardent 'believers' are, not surprisingly, the least open minded about the subject that they claim to be about 'the science'. Those who do not 'believe' that man-made emissions drive global warming because there is no proof of it are denigrated as deniers and sceptics and not to be given a voice. Science isn't silence.

Why is this so, as a scientist might ask?

One answer might be that the subject has morphed into a belief, a kind of secular religion (this is not my original idea) that provides ammunition for anti capitalists, socialists who want to change the world in their image and the world economic forum's many hopefuls. For them, it is a useful tool. (See elsewhere in this book for the IPCC official's admission that it's about the distribution of the world's resources.) That explains why it was brought into the public square with a predetermined outcome: the IPCC was given the mission to "address dangerous human-made climate change" ... The brief started with the conclusion it intended to prove, a classic circular argument. (See The IPCC chapter).

It's clear from the observed experience over the years that the adherents to the IPCC 'bible' (as it is called, ironically enough) fall into various categories:

- the activists, including some scientists and some in the media, who use the false narrative to change the world
- the naïve true believers, including some in the media, who have accepted the false narrative as 'gospel'
- the opportunists who seek to benefit financially from the manipulation of the market, eg subsidies for renewables to shut down fossil fuel production
- the political opportunists who don't really know the facts, can't be bothered finding out, just hoping to catch the voting wave

The public has not been able to rely on the mainstream media to scrutinise the alarming claims and those advancing them. Most media workers, like teachers, academics and 'thought leaders' joined the 'crusade'. Yet credible and factual information that disputes or calls into question the alarmism is readily available (eg the amount of carbon dioxide in our atmosphere). Even charities censor discussion that differs from the ruling orthodoxy by refusing donations (see DONATION CENSORED chapter).

The IPCC's charter fused natural variability, 4.5 billion years of historical cycles of warming and cooling, to human activity, and called it climate change.

That is tantamount to a criminal accusation without evidence accepted as a guilty verdict.

As Curry and Carter and other scientists point out, genuine scientific study of climate is ongoing – outside the IPCC's ambit, unheard and unseen by the public, by scientists who struggle with the wicked problems of understanding an "interdisciplinary subject requiring insights from many fields." Except the field of politics.

THE BOTTOM LINE

The question that haunts this subject is why alarmists prefer the doom scenarios in the face of evidence that humans are not causing global warming?

If policymakers (and responsible media) had not been fooled by the insistent but dishonest alarmism of environment and financially motivated activists, agenda-driven scientists and the IPCC, they might have scrutinized those claims of man-made catastrophic global warming before formulating policies that put all their trust in the clamour of 'WOLF!'

THE BOY WHO CRIED WOLF

There was once a young Shepherd Boy who tended his sheep at the foot of a mountain near a dark forest. It was rather lonely for him all day, so he thought up a plan by which he could get a little company and some excitement. He rushed down towards the village calling out "Wolf, Wolf," and the villagers came out to meet him, and some of them stopped with him for a considerable time. This pleased the boy so much that a few days afterwards he tried the same trick, and again the villagers came to his help. But shortly after this a Wolf actually did come out from the forest, and began to worry the sheep, and the boy of course cried out "Wolf, Wolf," still louder than before. But this time the villagers, who had been fooled twice before, thought the boy was again deceiving them, and nobody stirred to come to his help. So the Wolf made a good meal off the boy's flock, and when the boy complained, the wise man of the village said:

"A liar will not be believed, even when he speaks the truth."

Source: Aesop (1867) Aesop's Fables

DONATION CENSORED

The author offered to donate all royalties to Save the Children. This was the reply:

Dear Andrew,

Thank you for getting in touch with us and for offering Save the Children Australia royalties from your forthcoming book. While we appreciate the kind offer, from what we can ascertain from the description you have provided, the subject matter of the book conflicts with our donations policy as it appears to be spreading misinformation about the climate crisis. As such, we cannot accept donation of your royalties.

Save the Children Australia aligns its climate change programming and advocacy work with the findings of the Intergovernmental Panel on Climate Change, which provides the consensus of the vast majority of the world's climate scientists and lays out irrefutable evidence of the warming of our planet and the link between increasing, human driven, emissions of greenhouse gases, and increased warming and other significant impacts. [This is the basis of global climate alarm, unquestioned propaganda that has become the ruling orthodoxy.]

We appreciate your concern for the wellbeing of children, who, we agree, are the most affected by the impacts of the climate crisis. However, far from being passive victims, children are active agents of change in their communities, pushing governments to act faster and stronger on climate change. Save the Children Australia supports children's voices and advocates for increased action from governments globally. We also support programming that helps increase children's resilience in the face of unavoidable impacts.

While there are valid arguments around the best and most effective ways to reduce greenhouse emissions and build climate resilience, the facts of atmospheric science and the contributions human actions make to the changing climate are no longer up for debate. Save the Children Australia believes that it is time to move to implementing solutions at scale, not for debating the causes.

Paul Mitchell

Principal Climate Change Advisor | Save the Children Australia

At the time, a message at the top of the Save The Children website said **"URGENT: Children across the Horn of Africa are dying of hunger."**

The author offered to donate the royalties to *headspace*; the offer was declined:

"Unfortunately, headspace does not accept donations from the sale of books/literature without conducting a full clinical review as well as ensuring the messaging aligns with our brand. Ensuring our brand is safe for young people is our priority. We are grateful for your consideration to support the work we do supporting the mental health and wellbeing of young Australians.

Thank you for your understanding, and best of luck.

Kind regards,

Amy

amy coote MFIA

Head of Corporate Partnerships & Fundraising"

ADDENDUM

My first encounter with climate alarmism was on the roof of the Noga Hilton in Cannes during the 2006 Film Festival, which was crowning Al Gore in glory for his instant-gospel movie, *The Inconvenient Truth*. Naïve as I was at the time about climate alarmism, my interview faithfully repeated his claims about 'tipping points'and the apocalypse to come. Polar bears would die out. The polar caps would melt. Seas would rise in biblical proportions … Gore sent forth an army of believers, armed with copies of the film and a script to deliver in schools and community centres.

Within a few years as the global warming industry heated up, I gradually became uneasy at the proposition being trumpeted that fossil fuels were making the earth's temperature rise. I read everything I could on the subject. The more I read the more I resisted what was becoming the ruling orthodoxy, concerned at the amount of dishonesty that was being used to propel the narrative.

The following are just two of my articles (among several others on the subject) that were published in *Spectator Australia* that are precursors to this book.

Aussie scientist's 'consensus' report unethical
(*Spectator Australia*, 26/6/2017)

An Australia scientist's contribution to the global warming debate – the claim of a scientific consensus – has been described as 'unethical' by US research scholar Michelle Stirling^. She writes: "a claimed consensus is a powerful tool for driving policy but an inappropriate and unethical means of conducting scientific inquiry or informing the public."

Referring to the oft-cited report of a consensus by climate change communications, psychology and BSc researcher John Cook* (and others) Stirling says "Cook et al presents a collaborative work by several consensus study authors, who claim a 97% agreement by undefined climate science experts that "humans are causing recent global warming."

In a sober but stinging attack on the ethics of the Australian's work, Stirling writes; "The statement illustrates the problem of trying to use a social proof of consensus in place of scientifically defined evidence. The lack of empirical parameters that specifically identify the claimed ratio of human effect versus natural influence, the timescale in question, the level of risk or benefit, and the human activity or causative factor(s) are undefined. The notion of consensus defies the fundamental principle of scientific inquiry which is not about agreement, but rather a continuous search for understanding."

Stirling's paper, "97% Consequential Misperceptions: Ethics of Consensus on Global Warming", is published on the Social Science Research Network, a website established in 1994 and devoted to the dissemination of scholarly research in several fields; it is now part of the Elsevier publishing group. In January 2013, SSRN was ranked the top open-access repository in the world by Ranking Web of Repositories.

The Cook paper references the 2008 opinion survey of scientists by Professor Peter Doran and then-graduate student Margaret R.K. Zimmerman at the University of Illinois, Chicago, which claims a 97% consensus on global warming.

Sterling notes how "numerous earth scientist respondents … explained their view by email that the sun was the main direct and indirect driver of climate change, not humans or carbon dioxide. Indeed, as a result of that study author Zimmerman wrote: "I think I'm actually more neutral on the issue now than I was before I started this project." This important expression of uncertainty by a co-author of a consensus paper and these relevant views by earth scientists (a climate science discipline) are not mentioned, falsely inflating the cause and claim of consensus in Cook et al (2016)."

Further eroding the Cook paper's credibility are outdated references to consensus statements by national academies of science; "most of these were made before 2009, over a decade prior to the news of the IPCC reported their 2013 AR5 (Flato et al 2013) that there had been a 15-year hiatus in warming with temperature trends of "*values very close to zero*" (despite a significant rise in carbon dioxide concentration in that time). This is an important divergence of observations against modelled projections of the initial Anthropogenic Global Warming theory."

By failing to reference natural influence or these uncertainties "Cook et al (2016) creates a false and misleading public perception that humans are solely responsible for global warming/climate change, that fossil fuel use/ greenhouse gases are the sole factor, that humans can successfully stop global warming/climate change by reducing fossil fuel use, and that 'any' cost is acceptable to prevent a perceived danger."

The danger is not so much global warming but global misperception, says Stirling. "Efforts by Cook et al (2016) and others to establish a consensus statement on climate change for the public is improperly creating consequential misperceptions, as evidenced by the sweeping but inaccurate claims in President Obama's tweet: (May 2013): "ninety seven percent of scientists agree: #climate change is real, man-made and dangerous."

The misperceptions surrounding a 'consensus' are also consequential for society in terms of cost and psychological despair, Stirling claims, "and are particularly exploited by many Environmental Non-Governmental Organizations (ENGOs) as an effective means of fund-raising, relying on the claim that "...97% of scientists agree." With what, exactly, she asks.

"The failure to elucidate ... uncertainties and instead proclaim that there is an amorphous 'consensus on consensus' is an ethical breach," she argues, "and demonstrates a lack of scientific integrity. This is damaging to scientific inquiry, to science and society."

The last statement is worth some emphasis, especially in Australia in the immediate wake of the Finkel report on climate policy. Finkel is Australia's Chief Scientist, yet the report is not based strictly on science but a mix of politically accepted consensus and forward 'dreaming'. The scare tactics of exaggeration behind the consensus are copied from fairy tales to frighten children into good behaviour – or the boogie man will get you.

But there is a gold lining to the green cloud, collaterally for some large vested interests: Siemens, for example, has a US$250 billion slate for offshore wind turbine production over the next decade. They don't give a damn that the consensus is baloney.

Finkel, Malcolm Turnbull, Arnold Schwarzenegger and all of Hollywood, are not the only ones to have been conned by the claim of a consensus, which, as Stirling says, poses the question 'agree with what exactly?' It can be argued that Cook et al have done the most damage to the public discourse and to science, even more than Michael Mann and his misleading and discredited hockey stick graph and more than the East Anglia University Climate Research Unit's 'emailgate' scandal. It is Cook's 97% that is used as a hand grenade by the ill-informed (eg Dee Madigan on Sky News any time she is on a panel and the subject comes up) to detonate any discussion about global warming.

This fakery from zealots has been collectively adopted as driver of policy by the entire Western political class (the Chinese aren't that stupid). And we will pay for it in more ways than one. We have already got the electricity bill.

^ Michelle Stirling works for Friends of Science, which receives funding from the fossil fuel industries.

* John Cook is a Research Assistant Professor at the Centre for Climate Change Communication, George Mason University, Virginia. He holds a PhD in cognitive psychology at the University of Western Australia and a Bachelor of Science at the University of Queensland.

(The '97% consensus' claim was first debunked in Urban's article in the Spectator Australia of 15/10/2016.)

Parisites in our midst

(*Spectator Australia*, cover story, 5/5/2018)

Last month (April) it was two years since world leaders converged with self congratulatory ceremony and chilled champagne to sign the multi billion dollar Paris agreement on climate change. Next month (June) it will be a year since our Chief Scientist, Alan Finkel AO, admitted the brutish truth that, in effect, Australia expensively accommodating that agreement is pointless. Hello, Canberra?

Finkel was asked in June 2017, what difference would it make to the climate if Australia, with its carbon dioxide emissions less than 1.5% of the global total, somehow managed to stop emitting 100 per cent of its carbon dioxide.

"Virtually nothing," Finkel admitted.

He had just delivered his panel's Independent Review into the Future Security of the National Electricity Market. It had the word Independent glued in front of it as if to stick its tongue out to those who would mis-think of it as a report that toed any particular climate line. Except in its vision for the future, it lists four key outcomes, one being 'lower emissions'. In other words, the five authors (of whom Finkel is one) believe that emissions are too high (presumably of carbon dioxide, not cow flatulence) and contribute to global warming.

Make of that scientific contradiction what you will – and Finkel cannot be accused of being a dreaded 'denier'- but his view aligns with those who have been saying for a long time that our energy policies are hoodwinking the nation. Australia is an ongoing feast for the Parisites; these are the Paris agreement revering organisms that live off or within our national organism, obtaining nourishment and protection while offering no benefit in return.

Parisites come in ideological, political and financial form.

Ideological Parisites:

The well fed campus Parisite (Parisite Academus) claims the high moral authority to lecture their students and gullible commoners about global warming and climate change in the certainty that man's evil, selfish consumption of fossil fuels is scorching the earth, whipping up hurricanes and hastening us to Hades. There is no rational argument that can be had with this strain of Parisite, whose strenuous affirmation of climate change orthodoxy is a badge of virtue and righteousness to be protected with force if necessary. Their income is either directly or indirectly dependent on maintaining this position. Their campus minions can become thuggish …

The equally and obstinately misinformed public species (Parisite Plebeian) occupies positions in the bureaucracy and business where security of career requires total surrender to the ruling orthodoxy. They admire Al Gore. They are not curious. They are not sceptical. They are not well read on the subject. They confuse carbon with carbon dioxide. (They have an excuse; Finkel as a scientist, doesn't. He is taken to task for it by Dr. Michael Crawford: "The word 'clean' occurs about 50 times in your report, particularly in conjunction with what you label a 'Clean Energy Target'. It clearly implies that the alternative, in particular our fossil-fuel based legacy system … is dirty and thus ought to be replaced … you must be aware that the main emissions from fossil-fuel generators are water vapour and carbon dioxide (CO2) and not the element carbon in molecular or particulate form... (carbon dioxide) does nothing that fits with what people normally understand as "unclean" or "pollution".)

Political Parisites:

Easily the most pernicious of Parisites (Parisite Pox), the political strain is a mongrel of ignorance and incompetence, a genetic disaster. Such a Parisite is slave to groupthink pressure, feigns principles but shuns evidence based policy formulation, supports government largesse that appeals to Parasites Plebeian and Academus, lacks courage and fails the nation.

The most rabid and ego-fanning Parisite Pox (other than the Greens) resides within some State and local council bureaucracies, where they impose their will declaring renewable energy targets by post code, in defiance of the national plan. States that want to achieve the highest form of green purity opt for the highest levels of renewable targets. Imagine the benefit to our planet if only their State would aim to generate 50% or more of their energy needs with renewables. The renewables industry saw them coming. See below.

Financial Parisites:

Siemens is producing some US$200 billion worth of offshore wind turbines over the decade. Wealth Daily reported in December 2016 that Warren Buffett has doubled his stake in renewable energy, to US$30 billion. Australian renewable energy sources such as wind and solar will receive subsidies of up to $2.8 billion a year up to 2030.

'Renewables' is code for 'jackpot'.

The financial Parisites make up a wide range of sub-species in the Parisite Gimme group, ranging from large businesses in or around the many (renewable dollar) opportunities to home owners with solar panels subsidised by their neighbours.

Sadly, there is another, persistent and invasive Parisite Gimme sub-species and it gives me no pleasure to include it here. It is the vast army (army as in under orders and unified, saluting, marching, etc) of scientists and their assistants who toil at the coalface (as it were) of climate related sciences. I refer only to those who cling to the ruling orthodoxy about climate change, not the many who *"disagree about the causes and consequences of climate for several reasons. Climate is an interdisciplinary subject requiring insights from many fields. Very few scholars have mastery of more than one or two of these disciplines. Fundamental uncertainties arise from insufficient observational evidence, disagreements over how to interpret data, and how to set the parameters of models,"* as the authors of Why Scientists Disagree About Global Warming put it. (Craig Idso, Robert M. Carter (1942-2016), S. Fred Singer – November, 2015)

As these scientists also point out, *"climate scientists, like all humans, can be biased. Origins of bias include careerism, grant-seeking, political views, and confirmation bias."*

In my research I have not been able to discover grants, funding or subsidies enjoyed by scientists or scientific organisations tasked with finding the natural causes of climate variability. That research seems to be carried out in a self funded environment, free of Parisites.